What Color Socks Does God Wear?

Connecting our highest potential with our deepest fears

I0476543

By Doug Bennett

ISBN-13: 978-1512037616

ISBN-10: 1512037613

First edition, May, 2015

Acknowledgements

A book, like everything else in life, is the product of many influences, some known, many unknown. An ongoing influence are the Haden Summer Dream and Spirituality conference and the Natural Spirituality Regional Gathering conference. Bob Haden at the Haden conference and Heidi Simmonds at the NSRG keep inviting me back to give presentations. The groups listen to what I have to say and ask pointed questions and offer thoughtful comments. It is a experience of being accepted for me. Thank you.

My wife, Pat, has been patient with my endless writing when there are so many other things that I could be doing. She has been my editor and proof reader. I appreciate her contributions.

Contents

God Wears Socks

What color socks does God wear? Is that an irreverent question? There are many who think it is irreverent. Why would they think that? The obvious reason is that God does not wear socks. Only people wear socks, and people are very different from God. They are opposites, completely incompatible. God is immortal, all-powerful, all-knowing, and usually all good. People are quite mortal, very limited in their power and ignorant (compared to God), and certainly nowhere near all good. An old description is, "mortal sinners."

In the normal, conventional world view of our culture, our world has many things that are separate and different from one another, dualities, not related, or if we want to be very emphatic about the difference, we can say they are poles apart. If those things really are opposites and separate, it's a good thing to be aware of that difference. But if they aren't really separate, then believing they are separate or opposite and not related limits our experience of this life. It limits what we can expect from our lives. It limits what we can experience, what we can do, and as we will see, what we can feel. If, on the other hand, we are like God, or are of

God, or, perhaps, are God, then we should be able to do the kinds of things that God does. We should also be able to feel like God feels.

In our stereotypical cultural world view spirit and God are capable of much more than material beings. God and spirit can do all kinds of things I mentioned above and people cannot. Material beings and God are polar opposites. While this view is deeply entrenched in much of Western culture, many of the Eastern, mystical religions suggest that the separation between human and spirit is illusion and that the goal of spiritual practice is to find the unity that underlies the apparent separation.

It is my goal here to show that these two apparent polar opposites, God and humans, are the same, or are much closer to being the same than they are to being polar opposites. We are really wearing God's socks.

Our Quest

I am setting out to convince my readers that we material humans are really directly connected to the non-material intelligence of the cosmic quantum field and because of that connection we really can know what color socks God would wear, and we can wear them. If we believe that, the next question will be, why aren't we all floating around like happy Buddhas? I have an answer to that question, but will take us through the forest of feelings and, especially, the forest of fear. We'll finish up with some thoughts on finding our way through the forest so we can be happy Buddhas wearing God's socks.

We should probably expect to encounter some difficulties along the way. Here are some.

The Feeling Issue

I mentioned people feeling how god might feel in the previous section. In the interest of full disclosure I should say that I'm a guy, son of an engineer and a third-generation Scottish mother who wouldn't talk about her feelings if her life depended on it. I was raised as a card-carrying member of our patriarchal, science-based culture and I spent many of my formative years in engineering. The religion I learned about in my Presbyterian church couldn't possibly be right. It's not surprising that I might want to describe this underlying unity using science and reason and avoid the feeling issue entirely.

In my humble opinion I have had some success in using science and logic to explain spirit and the paranormal, which was very satisfying to me. However, it came as something of a surprise that my careful, logical thinking led me to the notion that feelings are important. They are really the center of our connection to the knowledge and power that is available to us through the intelligent quantum field. After I got over the surprise I found that introducing feelings allowed me to account for the process of connecting with the cosmic intelligence. It also provided a way to understand why, "What color socks does God wear?", is usually not a meaningful question.

The problem is that feelings are a problem for many people, not just guys. They are a problem even for people who are comfortable having and expressing feelings. Very few people are willing to examine their inner states. Even fewer people are willing to go looking around in their long buried fears.

The Non-Material Problem

We are setting out to unify God and human. One reason that God and science seem like they are poles apart in the world that we see and touch is that one pole is material and the other is non-material. Material things like you and me are local and temporal, that is they are only in one place at a time and they are only in one time, the present. At the other pole, the non-material "things" we know about can be in many places, or all places, at the same time, and they can be in all times at the same time. To put it another way, they can be everywhere and everywhen all at once. We all grew up in a material world. Science grew up in a material world. Our world is material. The material world is very normal and familiar. Non-material things are strange and alien. We'll see later that when things are outside our experience, it is very hard for us to even see them. Even when we do see them, as in the quantum experiments that demonstrate the non-local and non-temporal aspects of matter, we call them "quantum weirdness" experiments.

Now, having described non-material things as being far outside our normal experience, I want to point out that non-material, non-local and non-temporal "things" are actually a common part of human experience. That part of human experience is, of course, outside the conventional, mainstream, stereotypical Western culture. I am referring to psychics, intuitives, medical intuitives, dowsers, healers and all the others who know and do things without any apparent physical means. These phenomena are usually called paranormal. People who believe in the existence of God (non-material) are not usually included in the ranks of the paranormal, but they, too, have experience with non-material, non-local and non-temporal phenomena.

For many people non-material "things", like spirits, knowing and actions at a distance are not part of the "real" reality. They are imaginary, or fraudulent. That makes trying to combine the nominally real reality of science with the non-reality of the non-material a big problem. But that's our quest.

Forces Keeping Things Separate

The idea that God and humans are connected is not a new idea. The idea that there is not all that much difference between secular and spiritual has been proposed in many places. The mystics have always taught that God is inside each of us. There is the ever popular bumper sticker, All Is One, that summarizes that idea. There are even some lines in the Bible about everyone being of God, Ye are gods and all of you are children of the most high (Psalms 82:6). And the alchemist, Hermes Trismegistus, said, "Know ye not that ye are gods?" It has been known for a long time that the gulf between humans and God may not be as wide as it is often portrayed.

While it is an old idea, it has never been popular. There have been several factors throughout history that have encouraged belief that the dualism is real and large. Augustine and the early Christian ascetics promoted the idea that the material body was of little value and only the spiritual had value. Descartes got himself out of a bind with the Inquisition by proposing that the "transcendent mind" was not related to the material body. The church could have the transcendent mind, and science could have the material body, and the church wouldn't have to bother science anymore. Newtonian-Cartesian science has adhered scrupulously to Descartes's deal with the church and has carefully disallowed any non-material, transcendent ideas. Science is material, and that's that.

Over on the religious side it is important to churches that God be very different from mere mortals. It is important that God be big and powerful and good. If you said, "Come worship our wimpy little God. He tries hard, but don't expect much" you would have a very hard time filling pews or collection plates. People want their God to be strong when they need help. A lot of people have a strongly vested interest in having a God that is very much larger, stronger and different from us mere mortals.

In the sections that follow I am going to explain how the material and non-material parts of our world are really equal contributors to the makeup and functioning of the world we live in. That will serve as a good foundation for addressing the bigger issue of the sacred and secular divide.

The poles that we thought were far apart are not really all that far apart. Let's look at why they are not at all separate. We'll begin with the wave particle duality.

Bridging the Chasm

Science has come a long way since the end of the 19th century. There are now enough pieces available to allow us to talk about such esoteric things as the nature of God using the vocabulary of quantum mechanics, and some other things. As a sign of how far science has come, from its absolute, materialist, deterministic origins, science now has its own dualities; things that seem to be separate and incompatible but really need to be unified. I am referring to the wave-particle duality in quantum mechanics. Science used to be so comfortingly material.

Resolving the wave-particle duality will be a very useful guide to unifying our higher level polar opposites. Let's see how that works.

The Wave Particle Duality

Prior to 1900 and the beginnings of quantum mechanics, science had the world pretty well wrapped up. Just about everything had a neat and tidy material explanation. And those few things that hadn't been explained yet would surely be tidied up shortly. Of course, such a strictly materialist view of the universe had to ignore a great deal of non-material experience, but science at the time was on a roll and it could easily ignore those pesky details.

There were many things about quantum mechanics that were upsetting, but I would like to focus on one, that being the wave particle duality. Prior to 1900 matter was solid. There was some debate at the time about what the smallest piece of matter was, but matter was solid, no matter how small the smallest piece was. Some things were not solid. These included light and gravity. Light was considered by most to be a wave at this time. Although some, including Newton, thought that light was corpuscular, but light did so many wave-like things that most people considered it to be a wave.

Imagine the consternation, then, when Schrodinger's equation showed that solid matter is a wave. Protons and electrons and atoms were all waves when they were not being particles. By the time Schrodinger's equation came out, protons, electrons and atoms had been accepted as the appropriate model of material reality. Schrodinger showed that these little bits of solid matter were in fact waves in some thoroughly non-material medium, the quantum field. At first it was hoped that Schrodinger's waves were just an artifact of the math that was needed to correctly describe the behavior of atoms and electrons. But then experiments started to show that solid matter did behave like a distributed wave in real, observable, repeatable experiments.

One of those experiments, the double slit experiment, even showed that light behaved as a wave in part of the double slit experiment, but then became a particle when the light hit the film behind the double slits. Not only did particles of matter behave like waves, but waves behaved like particles.

It was bad enough that matter was a wave, but Schrodinger's equation also showed that these waves for small things were distributed across space and time. This is not how matter was supposed to work. Here we have the great wave particle duality of quantum mechanics. Our beloved, solid matter actually existed as non-material waves. These waves were distributed everywhere, and the waves were just probability waves. The wave functions gave all of the possible places that the solid particle might appear and the probability that it would appear in each place, but they did not tell us exactly where each particle would end up. Physics prefers equations that give exact, deterministic results. And now comes the worst part. Matter changed from being a wave-of-all-possibilities to being a solid particle, always a specific solid particle in a specific place. At some point the immaterial wave function collapsed into a very material particle. The details of that collapse were a complete mystery to physics. There was no way to predict the outcome of the collapse of a wave function. This was very embarrassing.

Here at the very core of physics, the description of how matter works, was this unexplainable duality. On a small scale there was no way to predict the outcome of a given event. This was very hard on a science that had long prided itself on being completely rational and entirely deterministic.

Unity of Waves and Particles in the Dance

Is there a way for us to understand this apparent duality as a single phenomenon, and if there is, how will it help us with our ultimate question about God's socks? I think there is a way to understand and it will help us with God's socks.

Schrödinger's equation tells us that everything that happens to a small piece of matter is described by the wave function. Actually, it's more than that. Everything that might possibly happen to a small piece of matter is described by the wave function. For a given piece of matter only a single thread of events actually occurs, but the wave function carries all the possibilities. If you want to know what's going to happen to a little piece of matter, you need to set up Schrödinger's equation and solve it for the wave function. The wave function tells us what the possibilities are for that piece of matter and how those possibilities change over time.

The normal description of the relationship between the wave and particle is that the wave function evolves until it is observed, or it interacts with something. Then the wave function collapses into a specific piece of matter at a specific place that we can't predict. The embarrassment for physics is that there appears to be no way to predict what the specific outcome of the collapse will be.

And that little description that I just gave is how the wave-particle relationship is usually described. The particle evolves as a wave and then every once in a while collapses into a particle. And that is the extent of the normal description. That is a description that I've read in many places over the years beginning with my quantum chemistry class back in college. That description makes it sound like there are indeed two very separate parts to the life of a piece of matter. One part of it evolves as a wave and then pop, it changes into a chunk of matter. That makes matter sound very, very different from the wave.

Roger Penrose, in *The Road To Reality,* Knopf, 2004, provided a more complete description of the relationship between the wave and the particle. It works like this. The particle does, indeed, evolve as a wave until it is observed or otherwise interacts with something. When it interacts with something, it collapses into a specific instance of that particle. But the rest of the story is that as soon as it collapses into a particle, then, poof, it turns back into a wave. After I read that, it was apparent it has to work that way. If the particle moves and changes in the wave function then it has to go back to the wave function after it collapses in order to keep moving.

Let's look at a material example. Consider a rock made out of silicon, aluminum and oxygen. That would be most of the rocks on earth. If I throw the rock, all of the atoms in that rock are moving. They move in the wave function and then they collapse into atoms, but the rock is still moving. In order to move the next little bit, the atoms have to go back to being wave functions so they can move and then collapse again a little bit further along their trajectory. What we think of as solid and permanent matter is really like an animation or a movie. It is a series of still images that change so quickly that we perceive them as continuous movement.

There is not a big divide in the life of an atom between its time in the wave function and it's time as a solid particle. The life of an atom is a continuous flickering back and forth between the wave and the solid forms. Since atoms can move very fast the time the particle spends in either form must be very short.

While the wave and particle forms are very different, they are intimately connected in that all matter continuously jumps back and forth between the wave and the particle forms. They are not really separate forms. They are two sides of the same thing.

This view of matter as being quite ethereal makes it easier for me to understand quantum mechanics and relativity. Einstein said that the mass of a piece of matter changes with its velocity. The length of a piece of matter also changes with its velocity. And matter itself can turn into energy. These ideas are difficult to understand if matter is a solid, permanent thing. It is much easier to understand if matter is this flickering, ephemeral state that never lasts very long. Matter is always jumping back into this non-material wave function.

The wave function determines what's possible for a piece of matter at any given time. The solid form of matter informs the next stage of the wave function about what is possible. The wave and the particle are in a continuous and very intimate dialog. They are not separate from one another in any way.

For our purposes, it is very interesting that the math tells us that for small particles these wave functions are non-local and non-temporal. Everything is made up of small particles. Everything, then, is potentially non-local and non-temporal. Of course most material things that we encounter every day are quite local and temporal. That big rock sticking out of the side of a mountain has not changed very much in hundreds of millions of years. How does that work? When small particles like atoms are in the close company of many other atoms, like atoms in a rock, the amount of time and the amount of change between collapses of the wave function is extremely short. Atoms in a rock cannot go very far before they interact with another atom so they appear to stay in one place.

This little meditation on waves and particles does a couple of things. First, it shows us that what we initially thought was a major duality is really a very fast dance between the two forms of matter. Second, it introduces non-locality and non-temporality into the world as described by hard science. Before quantum mechanics, non-local and non-temporal things were entirely in the domain of the spiritual, which was not at all real. We will need those non-local and non-temporal things to connect our material world to the spiritual.

We can "resolve" the wave particle duality by recognizing waves and particles are not a one-or-the-other issue. Matter is an intimate dance between the two forms. All matter is both, all the time. Gary Zukav described the S-matrix theory for describing subatomic particles in *The Dancing Wu Li Masters,* Bantam, 1979. He said of the dance among particles, "The dancers no longer stand apart as significant entities. In fact, the dancers are not even defined except in terms of each other. In S-matrix theory there is only the dance."(page 250) There is only the dance. The partners do not exist outside the dance.

Understanding the wave and particle forms of elementary particles as a single dance is interesting, but protons, electrons and pi mesons are not directly useful for describing humans, spirit and the relationship between them. We need human-level functions to do that. The idea of wave functions as a dance between wave and material forms will be quite useful.

Connecting People and the Field

Specifically, we need to understand how humans work in a world where the quantum wave function is the ground of all material being.

Our material bodies are both local and temporal, being made of matter. Our large scale, material bodies are clearly very different from anything that is non-material and non-temporal, like God. But our inner state is not material which makes it a good place to look for a connection with spirit. I have described how our inner state works in my previous book, *From Material Being to the Ground of All Being.* I will only give the very briefest description here.

Karl Pribram proposed that we think in holograms. That's interesting because the form of the quantum field, as given by the wave functions, is also very much like a hologram. Stuart Hameroff at University of Arizona has proposed that the micro-tubes in our cells, particularly in our neurons, are quantum computers that oscillate between a distributed quantum state and a collapsed material state. Hameroff has proposed that the micro-tubes are our organs of consciousness, which means that consciousness is a quantum process. I have extended that idea to propose that the micro-tubes transfer the information from the thought and feeling holograms in our brain to and from the quantum field.

What does it mean to be "in the quantum field"? As we understand it, the quantum field is wave functions. That means our thoughts/feelings have, or are, wave functions. Like wave functions for protons and electrons, that means that our thoughts/feelings originate in the field. All possibilities for our thoughts/feelings exist and evolve in the field. Every once in a while, about 10 times per second, a specific instance of conscious awareness collapses into our micro-tubes and we have a feeling or idea. I am using "conscious" here the way Hameroff uses it. It is not high-level human consciousness, but the intelligence of all life above a certain size. Actually, Hameroff and Roger Penrose calculated how big an organism had to be to support this process. They found that a starfish was big enough. Everything bigger than a starfish has this process of quantum consciousness.

That means that we have wave functions for our thoughts and feelings. It also tells us something about the nature of those waves in the quantum field. They are waves of thought and feeling, or consciousness if you like. Since our thoughts and feelings have no mass, their wave functions are distributed across space and time. If our thoughts and feelings are non-local and non-temporal while our bodies are quite local and temporal, then our thoughts and feelings exist before, while and after our material body exists.

I have mentioned that the form of the wave functions, and of the quantum field, is holographic. One of the properties of holograms is that they can carry vast numbers of apparently separate wave forms. Another property is that these wave forms resonate with similar wave forms and change each other.

With the ideas in the two previous paragraphs we can do what no other scientific theory has done (that I now of). We can account for the paranormal phenomena. Knowing and action at a distance, those anathemas in science, are actually as real as Newton's apple falling from the tree.

If we assume that the wave functions for thought and feeling perform like the wave functions for protons and electrons then we think and feel in the quantum field. All possibilities for all of our thoughts and feelings for all time exist in the field. The actual thoughts that we really have here in our material bodies represent a small and single thread of all the possibilities that exist in the field. Just like the wave functions for atoms, when the wave function for our thought/feeling collapses into a specific instance of a thought and feeling in our brains, it immediately returns to the wave function and informs the next set of possibilities. Just like atoms, our consciousness, our thoughts and our feelings are a continuous dance between the wave form and the material form.

What Is the Quantum Field?

Our thoughts and feelings are non-local, non-temporal and so they are immortal. And they are "in" the quantum field. When we talked about protons and electrons in the quantum field we said the field was waves, but we didn't know the medium of the waves. The waves, after all, just showed up in some equations that happened to have many sine and cosine terms. Now we have the proposal that the thought/feeling (feeling, mostly) of living things also has wave functions. This tells us something very important. The content of the field is thought and feeling, or consciousness. It is living, breathing consciousness.

That might be OK when we are talking about you and me, and maybe cats, who have some intelligence, but if the field "is" thought, am I saying that protons and electrons have thought, since they are "in" the field along with you and me? We don't normally think of atoms and electrons as having intelligence. To get a little ahead of my story, I will say that we'll see in the next sections that everything is an expression of the same single intelligence. What the intelligence can express depends on the limitations imposed by the physical form of the thing being expressed. Atoms and electrons just do not have a very rich physical form, compared to you, me and cats.

This is a very interesting development. With our thoughts and feelings distributed across time and space in the quantum field we can account for all sorts of paranormal phenomena and even the ideas of soul and heaven. This addresses the local, temporal versus the non-local, non-temporal part of the spiritual-secular duality. But I have suggested that we have some connection with God. We will consider God, next.

You, Me And God

The view of God that I will present here grew in several steps. The first step was built on the idea that the field is thought and consciousness and that each of us has a wave function for our own individual consciousness. If the field, itself, is consciousness, then it must be able to think on scales bigger than you and me. It was easy to imagine a big wave function in the field that would be the intelligence that people have always called God. I'm not sure what "big" means in this context. I assume that the wave function for the intelligence of a cat is smaller than the wave function for my intelligence. By analogy, then, we can think of the intelligence of God as being bigger than human intelligence.

That's interesting, but it makes God sound somewhat separate from you and me and cats. The second step toward the idea of God came from a rule in quantum mechanics. Roger Penrose described it in *The Road To Reality,* page 578, Knopf, *2004* . The rule is that when two things interact they are both included in the same wave function. In conventional physics if you have two things and you want to describe what they do, you write one equation for each. If they interact with one another then you write a third equation to describe the interaction. Quantum mechanics is much messier. When things interact they all get included in one wave function, which means wave functions get complex very quickly. The result of this rule is that the wave function even for a single atom cannot be calculated exactly. All of the protons, neutrons, and electrons in the atom are interacting so they are all included in the single wave function for the atom. We don't know how to solve for that kind of wave function.

I interact with people around me, with my cat, with the people I meet on the street. And all those people interact with lots of other people and their pets and even the mouse living in the basement. Actually, everything in the universe interacted with everything else back in the Big Bang. That means that there is only one wave function in the universe. It includes the intelligence and the animation of everything there is. When I first put those words together the phrase, ground of all being, jumped into my head. I had just described something that is, quite literally, the ground of all being. I did not invent that phrase, of course. That phrase has been used to describe God for a very long time.

The Ultimate Connection

God, in this view, is inclusive. Everything in the universe, from quarks and electrons up to you and me, and everything else, is an expression of the intelligence that is the cosmic wave function. Everything in the universe is in a flickering dance between the non-material wave form and the concrete, material form. What happens in the material world is driven directly by the intelligence of the universe when the wave functions collapse into specific forms in the material world. The states of the wave functions are altered when those collapsed material forms return to wave functions and inform the next set of possibilities.

What does this tell us about our view of the secular and the spiritual as polar opposites? It tells us that our life in the material world and as material beings is not separate from the spiritual world of God. We are all equal expressions of that single intelligence and our choices and actions alter and inform that single intelligence. We are both expressions of God and contributors to the intelligence of God.

I think that is an amazing idea. It validates key teachings of many mystical traditions. It makes the All Is One bumper sticker a statement of literal truth. When Hermes Trismegistus said, "know ye not that ye are gods?", he was right on.

God is not separate from us. God is us, or, at least as much us as our material form and learned experiences (more on that later) allow us to express.

If we humans are somehow expressions of, and part of, the quantum consciousness that is God, is it possible for us to be like God in the material world? Is it possible for us to live like we are expressions of the consciousness of the universe? If we did live like that, would it be different from the way we live now? That is, are the socks we wear, God's socks?

Human Socks, God's Socks

If we are connected to the ultimate ground of all being, indeed, if we are direct expressions of that cosmic consciousness, then we should be able to live like that was true. We know that is possible because a few people have always lived like they were direct expressions of God. Just what does a direct expression of God look like?

Acting Like God

Because the single intelligence of the field is waves, and holographic, that means that connecting with things beyond ourselves is entirely possible. We can know things at a distance. Since the field is all non-material waves then we can influence the things that we can connect with. We can connect with things that are non-material, like the thoughts, feelings and health of other living things. These kinds of activities are usually called paranormal activities when we talk about people doing them. But they are also similar to behaviors attributed to God. God can know things and create things and alter the outcome of things in the material world. In that sense we are gods, but on a scale that is limited by the current state of our material being.

Feeling Like God

And because our consciousness is an integral part of the larger consciousness that is the ground of all being, then it seems that we should be able to feel like God feels. How can we know what God feels, you might reasonably ask?

Let's go back to electrons. The wave function for an electron is a single, unified holistic entity. No matter how non-local it is, no matter how much of our material space it is spread over, anything that happens at any place in the wave function instantly affects the entire function. It's not like a water wave rolling up the beach and hitting your ankles. The part of the wave that hits your ankles is disrupted, but the rest of the wave continues on up the beach. If the water wave were a wave function and it rolled up the beach and hit your ankles the whole wave would instantly collapse. I take this little example from particle physics to mean that any intelligence that includes many other intelligences, like the single wave function that is the consciousness of the quantum field, is aware of everything. It is aware of what's happening in the material world and it must be aware of all the possibilities that exist in the field, but that don't actually happen in the material world.

Another way to say that is that the intelligence of the field is infinite awareness. And since it is aware of everything that is and everything that might be, there are not likely to be any feelings of, "I wish something would be better," or, "that shouldn't happen like that." In other words, intelligence of the field is also infinitely accepting, not because it's a virtuous way to be, but because that is the only way to be when you can see all that is and all that might be.

How does God feel? I believe God is complete awareness and acceptance and discernment. Here in the material world when someone achieves that state we say they are enlightened. One of the hallmarks of enlightened people is that they are serene. God, then, feels serene. Or, God is serenity.

Since we are part of the intelligence that is God, it should be possible for us to feel like God, to feel serenity. And, there have always been a few people who have achieved that state. Those people are widely admired. It is a pleasure to be in their presence.

Our Highest Potential

Our material bodies remain resolutely local, temporal and material. But our inner life is intimately connected to, is even part of, the ground of all being. We are not separate from God. The duality comes from viewing humans as only material beings.

We will see later that our physical bodies play an essential role in our perception of the non-material intelligence. We can think of the non-material quantum consciousness as having many feet in the material world. And most of those feet wear socks.

Potential is an awful word. When other people think you have potential for something it places a large and unreasonable burden on you. The potential that other people see in you may or may not have any bearing on the possibilities that really mean something to you. Having said that, it is clear from this little meditation that it really is possible for us to live like Gods. We all have that potential. And we know that is true because there have always been a few people who have actually done that. That makes it possible for all of us to do that. We can live like we are expressions of the highest intelligence in the universe. Since we live in this material world with material bodies and feet, we need to wear socks. The question of what color socks does God wear is not a trivial question. A serene and enlightened God would probably wear serene and enlightened socks.

What sorts of things are possible? What is the potential? Michael Murphy has been a student of human potential for most of his life (he was born in 1930). He summarized the results of that study in his book, *The Future of the Body*,

Tarcher, 1992. In every walk of life, in every era, there have been people who have transcended the normal bounds of performance. From observation we know that it is possible to be more than average humans. From our meditations and musings in this document we know why those things are possible.

We do have amazing potential. We can know what color socks God would wear. But another observation is that the people in Murphy's book have always been a very, very small fraction of the population. Why is that?

But What Color?

After all that careful reasoning about the nature
of God and our potential to be expressions of
God, if we look around we quickly see that the
potential is a distant potential for most people.
The great majority of us are not floating around
like serene and happy Buddhas. In fact a great
many of us live lives that are full of quiet
desperation, or frustration, or depression, or
pain and suffering. To make matters worse, the
great majority of people whose lives are not
happy and serene, vehemently resist any
attempts to change their state of being. They
spend their lives feeling bad and believe that is
their lot in life.

The fallout from feeling bad goes beyond
individual lives. Our inner states, good or bad,
are part of the consciousness of the quantum
field. If our inner state if filled with desperation,
frustration and suffering, then that is what we
are contributing to the cosmic consciousness.

Why We Aren't All Skinny, Rich And Happy

While many people resist all attempts at inner change, there is another fraction of our population that is interested in change and transformation. We are called the self-help nation, because so many of us buy all those books on change and transformation and healing. We go to study groups, we go to seminars and workshops and we buy lots of books. It is a large industry. The title of a book by Brian Klemmer sums up the situation nicely: *If The How To's Were Enough We Would All Be Skinny Rich And Happy,* Vision Imprints Publishing, 2005. In spite of all this interest and work directed at personal transformation, I have not seen much evidence for any wave of transformation and enlightenment sweeping the country. There is some change going on, but the pace is slow. There is still much suffering and even when people want to become healthier or more serene, or whole, it appears to be very difficult.

Why do we not live our god-like potential? And even when we want to learn how to do that, why is it so hard to achieve? The answer is that we all learn our own realities here in the material world. And what we learn is usually something other than "we are all of God." Our normal lessons usually include things like, "we are material beings", "we cannot be anything like God", and Augustine's idea about original sin is still popular in some circles.

The ability to learn our realities based on whatever situation we find ourselves is a major contributor to humans' ability to dominate life on earth (or at least our ability to think we dominate life on earth). We can learn to survive and to thrive everywhere, from tropical jungles to Arctic tundra, from seashore to mountains. This is a very valuable skill. It is also, as we will see, the cause of much of our pain and suffering.

This is where the duality about God being different from humans came from (in part). The great majority of humans do not live lives that are very god-like, in the normal sense of "god-like". This is where we encounter fear. This is where we connect the lowest of human emotions, fear, with our highest potential. Let's begin by looking at how we learn our realities.

Learned Realities

How do we learn our reality? Isn't reality just what is "out there", the stuff that we walk around in and bump into once in a while? Let's look at how that works.

We perceive our environment, what we think of as reality, with our senses. Pribram proposed that the process we, and all other living things, use to perceive things involves memory of learned experience. In Pribram's model we take in information from our environment through our senses and convert it to holograms. Those little holograms are taken to our memory, where we retrieve something from our memory that matches the input information. That retrieved information is projected back out of the receiving sensory organ and forms what we see or otherwise perceive.

This means that everything we perceive is learned. We can obviously learn to see all kinds of things, but it takes some work. My favorite example is serious birdwatchers. Serious birders see and hear all kinds of birds that people who have not taken the time to study bird watching simply do not see. At the lowest level of physical, sensory perception, everything that we perceive is learned. It comes from our past experience.

There is more, of course to our realities than just recognizing the presence of things. We can recognize situations, for example. We might recognize a large room filled with chairs all facing a stage with a podium on it as the physical arrangement of things, but we might also recognize that situation as one where we have to give a talk to a large group of people. Different people will have very different reactions to that same physical situation. Some people will be pleased at the prospect of giving a talk to people. Many others will be terrified at the prospect of speaking before a group. Our perception of the physical objects also carries meaning and feeling reactions. I want to propose that the meaning and feeling reactions occur at the very same time as the lowest level, physical perception.

The example I like to use for this is snakes. People have a wide variety of reactions to snakes. I rather like snakes, I find them very interesting. I am, however, afraid of stepping on them. If I am walking through high grass and notice something curvy near my feet out of the corner of my eye, my first reaction is to jump away from it. The physical reaction is the first thing that happens, before I recognize that there might be a snake on the ground. This is a learned reaction. Someplace in my early life I learned that stepping on a poisonous snake is dangerous, so I am afraid of doing that today. This seems to have been a rather "strong" lesson because throughout my life it has manifested itself as an immediate physical movement, jumping back from the possible snake on the ground. The important point here is that my reaction to this situation involves not just sense perception, but an immediate physical response that was learned.

Packaging our Learning Experiences

Perceiving our realities is a rich and nuanced experience for each of us. I have introduced a couple aspects of our our perception, physical recognition and physical action. There is much more involved in our perceiving of reality. Later, we are going to talk about changing how we perceive reality and so it is important that we understand what all the parts are. Here, I would like to fill in the other parts of how we experience our environment. It is important to keep in mind that all of these parts are learned in our initial experience with them.

There are several, apparently different aspects to our learned responses. To give them all a convenient handle I have invented something I call a perception packet. I am not proposing that such things really exist. This is just a way to explain the process. Imagine that for each situation in our life we have a little packet, a perception packet. Every time we encounter that situation or that person or thing we look up the packet for that situation, open it up and it tells us what we have encountered, how to feel, and what to do in that situation.

Earlier, I mentioned feelings as being part of our learned response. I'm sure there are many people out there who believe that feelings are an important part of everything. But for those of us (guys, mostly) who learned that feelings are not important, some justification is needed. Later I will describe the role that feelings play in our learned realities. For now I can offer a justification for the role of feelings based on brain function. The amygdala in the limbic system (the mammal brain) connects feelings and action. The limbic system developed before humans and their cerebral cortexes. Non-human mammals only have feelings. They have no words or conscious thoughts. The amygdala serves to connect the learned feelings with the learned actions. It still performs that role in humans. Its input is still only feelings, especially fear, and its output is movement and action. No words or high level interpretation is required.

The feelings that we have in our initial encounters with a situation have a big impact on what we learn. If we are afraid in a situation, we are going to learn something very different compared to what we would learn if we felt good or happy in a situation.

The high level feelings we have in a learning situation are very important in shaping our response to the situation and what goes into our perception packet. That initial learned feeling persists in our perception packets because they are used as input to the amygdala to produce our initial, physical response in the situation.

Thus far, our perception packet contains the physical recognition of the object or situation, a meaning or interpretation (in words) of the situation, a feeling and movement or physical reaction. There is one more component in our perception packet: the felt sense. This term was coined by Eugene Gendlin. Gendlin was a psychology researcher at the University of Chicago. He was looking for factors that could be used to predict the outcome of psychotherapy. After listening to recordings of the complete course of therapy for 124 people he found a factor that successfully predicted whether the therapy would be successful. People who referred to, or checked in with, sensations in their bodies during therapy had successful outcomes. Those who did not refer to body sensations, did not. He called these body sensations, the felt sense. The felt sense precedes our reactions to all situations (by a very short interval). According to Gendlin, the felt sense is how we know that we feel happy or afraid. It is how we know that we have forgotten our keys as we walk out to get in the car. It is also how we know if something "feels right".

The felt sense is a physical sensation in the body. It is different from the higher level meanings or emotions that might be associated with our response. Since it is part of our response to all situations, it is part of all of our perception packets. It is important to our understanding of our learned response to situations because, as Gendlin showed, our conscious awareness of the felt sense is critical to our ability to change and to develop new perception packets.

Our learned perception packets, and, thus, our learned realities, contain physical recognition, a felt sense, higher level feelings and emotions, actions, movement and behaviors, and meanings and interpretations. All of these elements are learned in our initial contacts with the situation.

Living With Our Lessons

As a result of this basic perception mechanism, we can say that reality, the reality that each of us perceives as an individual, is learned. That means that we can each learn a different reality. We do a very good job at learning a shared reality for many of the material aspects of our world. That's not too surprising because everyone's experience with bumping into walls and touching hot stoves is very much the same. On the other hand, the meaning that a situation has and the kind of behavior that we exhibit in the situation depends very strongly on the initial feelings that we had in that situation and those vary widely. The result is that individuals have very different reactions and behaviors around snakes and public speaking, for example.

How we perceive and behave in a given situation can change over time, and many people do experience a change in the way they perceive a given situation. But it is also possible, and quite common, for people to carry their original learned reactions and feelings about a situation to the grave.

Feelings, Again

Our reality is determined by the feelings that we had when we first encountered each aspect of our reality. The initial feeling that we had during that learning experience is usually the one that persists throughout our lives. We can say that feelings drive our perception of reality and so they also drive our habitual behaviors and our responses to our reality.

This, I believe, is the major reason why we are not all happy Buddhas. Very few of us had early learning experiences that were anything like "being God".

Thus far I have made our response to our environment sound rather passive. It is a perception mechanism. I have also said that very few of us live "god-like" lives. But our response to our environment is very active. We play a very god-like role in creating the reality we live in and the life that we each lead. It doesn't often look like we have god-like powers of creation because the lives many of us lead are full of trouble, pain and suffering. And everyone knows that a right and proper god does not create pain and suffering. If you look around you will see that each person does an excellent job of creating the reality that they learned. It may not be pretty, but it is exactly what each person learned.

We create our realities with the choices we make and the actions that result from those choices. Because of our connection to the larger intelligence of the field, we also influence what happens to us and the people that we encounter.

We actively create the reality that we learned. Even if it does not look like our individual realities were created by gods, they were, and we are those gods.

Real Reality and Feeling Right

If all of our reality is learned, and we can learn good things and bad things, then how do we know what's really real? The answer is that we can't know, in any concrete sense, what is real because of the nature of our perception mechanism. In looking at how people operate and noticing the difference between people who feel good in their lives and people who do not, (that is the only measure of "success") I conclude that the very best we can do in perceiving the truth of our reality is to rely on what "feels right". When I first thought of that I unconsciously equated feeling "right" with feeling "good". Then I had a problem with short-term, frivolous feeling good and the more sincere forms of feeling good. But I don't think that "feeling right" and "feeling good" are the same thing. I believe that the very best we can do determining what's really "out there" is to check in with our body and see what feels right. I say that because people who consistently do that seem to feel very good (quite sincerely).

How Many Feelings?

I said that the feelings we had in the initial experience determine what we learned in that situation. There are lots of feelings out there, from low level body sensations through all the emotions up to meanings and values. It looks like a complicated process. The process can be simplified a great deal with an idea that the philosopher, Krishnamurti proposed. There are only two emotions: fear and love. Everything else is combinations of those two basic emotions. I don't know if that is true and I am sure that many philosophers and therapists would disagree with the idea, but it is very useful for our purposes here. It is also a good fit with the idea that our consciousness emerges from the larger consciousness of the quantum field. Let me offer a slight digression on the nature of love and fear in the quantum mind.

Love and Fear in the Quantum Field

Love has been an important value, if not the most important value in spiritual teachings for a very long time. It can influence all aspects of our lives for the better. Is there a way to understand love as coming from the quantum intelligence of the universe? Early on in this work I had wanted there to be some greater value placed on goodness in the quantum field. It was a personal bias that I did not want the quantum intelligence to be completely devoid of values. I was very happy when I found a way to connect the quantum intelligence with what we call love.

The quantum intelligence runs on connection. Everything is a living, vibrating hologram, and a single hologram at that. This means that any two things that have any sort of resonance with each other at all, can, and do interact. Everything is connected in the quantum field. This connection is not an arbitrary virtue, it is simply the way things are. Here in the material world, where the connection between things is not immediately obvious, we value connection when it does happen. We feel better, we are healthier, and we can survive infancy and childhood, if we are connected with others. This feeling of connection is not required however. We can be isolated and we can isolate ourselves from connection with others. We have given a name to connection, actually several names. One of them is love. Others include forgiveness, acceptance and compassion. The value that we place on love is a reflection of the basic state of the quantum field intelligence, that is, it is a reflection of the basic state of God. One way to say that is that the field is connection. Another way to say it is that God is love.

And now I can add Krishnamurti's idea about emotions. There are only two emotions, love and fear. Love is connection. Fear is isolation, separation. Those two states characterize the way we can be in the material world, and in our relationship to the larger intelligence of the field. Connection and isolation are the only two ways we can relate to the field intelligence. It is reasonable to say that there are only two emotions here in the material world, love and fear.

If there are only two emotions, or feelings, then we can conveniently divide all of our learning experiences into those where the feeling was love, and those where the feeling was fear. In the experiences driven by love, we learned connection. In the experiences driven by fear, we learned separation and isolation.

Fear-Based Realities

Fear-based perception packets affect our lives in several important ways. They all separate us from knowing what socks God would wear. All of our perceptions about our reality are learned. What we learn in any given learning experience depends on the feeling that is associated with that experience. My generalization is that when the feeling associated with the learning experience is fear, we develop a distorted view of reality. "Distorted view of reality" makes it sound like there is a "true" reality out there to distort. That is not the case. Here, distorted, means "based on separation and isolation."

Fear, itself, is not a bad thing. It is a very important warning system that tells us about situations that are dangerous. Fear in a dangerous situation is normal, healthy and very useful. The kind of fear that I am talking about is fear that is associated with our learned perceptions of reality. It is a fear that might have been real and appropriate in the initial learning experience. It may also have been a fear that came from the inaccurate interpretation of the situation by a young child. It becomes a problem when our response remains fear-driven when the source of the fear no longer exists. Then the learned experience no longer serves us. This learned fear causes us to avoid situations that could be very health giving for us. Examples are when we learn to be afraid of close relationships, or public speaking.

A learned, and now inappropriate, fear is a red flag that indicates that we are viewing our reality through a distorted lens. A fascinating result of learning all of our perceptions of reality is that there is nothing "absolute" about any of our experience of reality. The only way we can know what's "right" or "true" is with context and intuitive checks.

A fear-based learning experience produces other effects in our lives. Peter Levine, in his book, *Waking the Tiger,* North Atlantic Books, 1997, says that the unresolved trauma experience, which is fear-based, causes serious distortions to our realities. An unresolved traumatic experience involves a large fear: death is imminent. If the energy of the experience is not released (resolved), the body learns to repeat the terrible experience in order to try to resolve the trauma. The reality of people carrying an unresolved traumatic experience is seriously and dangerously altered so that they will willingly re-create the experience, hoping that it will be resolved this time. And even though the experience is dangerous and has negative outcomes, people will repeat the experience many times. Further, they will often resist any attempts to heal themselves.

Levine notes another effect of unresolved trauma. It causes us to lose our connection with our body sensations. We stop being aware of, and stop trusting, signals that we get from our body. We can be cut off from our body sensations by many kinds of life experiences other than trauma. This fear masks our awareness of body sensations. The result is that we have a limited ability to "feel what's right". Fear both distorts our view of reality and makes it difficult for us to determine that we are looking at the world through a distorted lens.

Connecting with the Field

On one hand, we are always connected to the field. There is no other way for us to be. On the other hand, it is possible and very common for humans to believe and behave as if they were not connected with cosmic consciousness.

Our connection to the content of the field, both sending and receiving, is driven by our feeling state. Consider long term memory. Long term memory is not located in the physical brain. A good candidate for it's location is the intelligence of the field. It appears that in order to move memories from short-term memory into long-term memory we need some sort of feeling or emotional connection with the information. Technically, the vagus nerve needs to be activated in order to move information into long-term memory. The vagus nerve connects physical sensation in much of the body to the brain. That's why it is usually easy to remember things that were associated with strong feelings and is difficult to remember things that did not have a strong feeling associated with them.

Energy healing is an example of influence that we exert through the medium of the field. The universal advice in energy healing classes is that it is very important to be in a relaxed and centered state of being when doing healing work. Being relaxed and centered is a state of feeling. Exerting influence through the field requires a particular feeling state in the practitioner.

Receiving "new" information from the field also requires a particular feeling state. The required state is actually the same feeling state required for healing work: being relaxed and centered. When we are working on a problem or having some difficulty it is quite common to work and work with the problem and then give up. Then a little later, when we are waking up, or going to sleep, or taking a shower, or walking in the woods, a solution or an answer pops into our head. That, I believe, is information received from the intelligence of the field and it requires a particular relaxed feeling state on the part of the receiver.

I conclude from arguments like this that it is feelings that drive our communication with the field. It is not our conscious minds or conscious will. The next step is that the intelligence of the field is what we here in the material world call feelings. To receive inputs from the field, we have to be able to perceive abstract, body sensations. How does that work?

Content of the Field

The content of the intelligence of the field, is what we call feelings. I arrived at this idea by thinking about composers, specifically Mozart and Sousa. Mozart was a prolific composer and, by his own description, did not actually compose his music. He just wrote it down as it flowed through him. Sousa wrote marches and at least one of them came to him in a dream. He dreamed he was on a cruise ship and there was a brass band on deck playing a march. He listened to it and when he woke up he wrote down what he heard. It came to him as a complete composition. I believe that both of these examples are instances of people receiving information from the larger intelligence of the field. The question is, does the field compose music for Mozart and arrange it for an 18th-century string orchestra and then compose music for Sousa and arrange it for an early 20th century brass band? I suspect that the field does not make that distinction.

Instead I believe that the field supplies us with abstract forms, structures and relationships. This abstract information is relevant to whatever it is we are working on, but we have to use our learned information in our perception mechanism to wrap the abstract information in concrete forms that are relevant to our specific problems. Mozart got a feeling that carried abstract information about forms and relationships. He used his learned music vocabulary, that of an 18th-century string orchestra, to wrap that abstract information in the concrete form that he wrote down. Sousa used his learned musical vocabulary, associated with a brass band, to wrap the abstract information that he received in brass band music.

These abstract forms are not specific ideas with specific forms. Being abstract, we perceive them, initially, as "abstract" sensations in the body. I have used the term, perception process, to describe connecting abstract information with specific forms. I have also used the term, "perception packet", to describe the learned connection between learned feelings and specific meanings, feelings and actions. The idea of a perception packet emphasizes the arbitrary nature of the connection between, sensory (or non-sensory) input, the feelings we have and the response we learn.

Why did Mozart get abstract information that related so well to 18th Century orchestral music? The answer is that he was having feelings about writing music. He was probably not "thinking" very much about it. He was just in the flow of music. In terms of my model of the intelligent hologram, he was putting feelings about music into the field and those feelings resonated with related information and he connected with that information. Another, less technical, way to say that is that he "attracted" it because that was what he was "putting out". Because the intelligence of the quantum field is a resonating hologram, everything about our individual inner state is resonant with everything that is related in any way to our current state. Since the field is the source of everything, then ideas, people and events that appear in our lives do so because they are resonant with some aspect of our inner state, specifically, our inner feeling state.

Fear and Separation in the Material World

We have seen that fear and isolation affect our local lives. They have a larger role, as well. Fear and isolation limit our connection and communication with the intelligence of the field, that is, with God. In other words, it is very difficult to know what color socks God would wear if we are looking at the world through the lens of fear. Let's consider how that works.

The fear I am talking about here is not the feeling that arises when one is confronted with a truly dangerous situation. The fear I am speaking about here is the feeling we harbor about situations where we believed we were isolated or disconnected in our original encounter with the situation.

Fear as isolation is an interesting concept. We cannot be separated or isolated from the intelligence of the field. Everything we do is a direct expression of that intelligence. That is the nature of existence. We can, however, be isolated from our own feelings and inner state, and we can be separated and isolated from connection with people who are important in our lives. I'll call this "local isolation" to distinguish it from isolation from the ground of all being.

Even though we cannot be isolated from the intelligence of the field, we can interpret the connection through the lens of fear. Feelings of local isolation have a profound impact on our connection to the cosmic field. Our contribution to the growth of the field intelligence is our inner feeling state. If that state is dominated by feelings of fear and isolation, then we are contributing fear and isolation to the field. That's not bad, of course (we wouldn't want to be judgmental) and there are a great many people who are contributing fear and isolation to the field. But it seems to the responsible first child that I am that I should be contributing something good rather than something not so good.

Putting out feelings of fear and separation has a more practical result. In the world of quantum intelligence, resonating with something is the same as connecting to it, influencing it and bringing it into our lives. Just as Mozart "put out" feelings about music and received music-related information, if we "put out" feelings of fear and isolation, we will receive information related to fear and isolation. The phrase, "put out", makes it sound like that is something that we do explicitly, as in, "Well, now I'm going to send out some fear to the universe." That is not the way it works. Our inner states, whether fear or an intent to make music, only exist in the field. We don't have to "send" it anywhere. It is always out there mixing, resonating, attracting and influencing everything. The result is that if a significant amount of our learned realities are fear based, we will find our lives filled with things that support our fears. That's not to say that is good or bad, it's just the way the universe works.

While we cannot be isolated, or separated from the intelligence of the field, no matter what we believe, we can be separated from other people and from our own inner state. This separation has well known physical and psychological costs. The cost we are concerned with here is that fear isolates us from our own body sensations. This is critical because being aware of the subtle body sensation that answers the question, "Does this feel right?", is our only check on what is real. I find myself using the term, "real", as if there was some objective reality out there. There can be nothing "objective" out there when all of our perceptions and interpretations of the world around us are learned.

The clear implication of all this for me is that addressing our learned fears is the most important part of any inner work that we might undertake. I believe that learned fears lie at the root of many spiritual and emotional problems, of many chronic physical problems, the "bad luck" that seems to dominate people's lives, and even difficulties we have in learning new skills.

A Science of Spirit?

We started out trying to bring the polar opposites of the secular and the spiritual together. Science, today, is nothing, if not secular (but changing as we speak). Everyone is familiar with the apparent conflict between science and religion. If "religion" means organized religions, then I think the conflict is real and in many cases entirely justified. I have tried to show here that the material and non-material worlds, or the sacred and secular worlds, are really partners in a very intimate dance, so intimate that the partners don't exist. There is just the dance.

There is another way to bridge the apparent gap between sacred and secular. It is to make the spiritual domain available to scientific investigation. Such investigation would be different from the current, mainstream scientific investigation. But science is supposed to describe our universe. If spirit is a "real", but not material, part of our universe, then it should be fair game for a more enlightened and inclusive science. The science musings that I have described here and in my other books are an attempt to do that.

If the science that would investigate spiritual issues is different from the current, mainstream science, then the spirit that science would study is different from the spirit described by most, mainstream religious organizations. I am quite confident that there will never be an equation whose solution begins with "Thou shall not ...". What does the spirituality look like that would be part of a (new) science-based reality? Many of the old ideas are supported in the "quantum religion" but with some differences from the normal usage.

I said that the single wave function of the quantum field is what people have always called God. We said that our consciousness is distributed across space and time while our material bodies are very local and temporal. That means that our consciousness exists before our material body appears, it animates the material body during its life here in the material world, and it continues to exist after the material body stops existing. That is an excellent description of what people have always called soul. Our soul is our wave function. Our personal wave functions are part of the single wave function that is the universe, or what people have always called God. The conventional view is that God and souls live in heaven. In this quantum view, the single, distributed wave function that is the consciousness of the field is what people have always called God and souls. That means that the quantum field is also heaven. God, heaven and all the souls are all made of the same thing, the quantum field.

It appears to me that if the wave function of our consciousness persists across time then the possibility of having multiple lives (material bodies) is reasonable. And if we have multiple

lives with the same wave function –
consciousness – then that sounds like another
old spiritual idea: karma.

In conventional religious terms, that means that
we are all in heaven all the time. It's where we
live. There is no other place to be. But what of
that other part of conventional religion, hell? In
the model described here, hell is the belief that
we are not connected to god, or that we are not
in heaven. That belief, like all our beliefs is
something we learned. Our learned beliefs define
our reality while our material bodies exist, and,
if our personal wave functions persist across
multiple instances of our material bodies, then
our beliefs should persist as well. We learn how
to make our lives here in the material world and
what we learn persists between and across our
material lives. We make our own hells, heavens
and everything in between.

As the mystics have always said, God is within
each of us. Even the bible says, " for, behold,
the kingdom of God is within you. ", Luke,
17:21, King James version. We are in heaven
now and heaven and God are all within each of
us. That strikes me as a much more believable
and desirable reality than having God outside us
and having to obey all kinds of arbitrary rules

specified by an organization to be allowed to connect with God.

All that sounds quite rosy. But on average the human condition is much different from the rosy potential. Most of us do not live lives that feel connected because we learn our realities (a very useful survival skill), but most of us learn that we are not connected. When we are not connected we feel fear and we greatly limit what we believe is possible in our lives. Buddha noticed the same thing once he got outside the palace walls.

For those who are not concerned with making things different, either because things are good the way they are, or because they believe that things are supposed to be the unpleasant way they are, nothing further is required. You can stop reading here.

For those who are seeking something different and better, or those wanting to feel better, or experience a closer connection to the divine, or those who want to fix all your faults, read on. If all my preceding discussion about how things work is anywhere near accurate, then we have a good basis for developing practices that can help move us toward a rosier reality.

Wearing God's Socks

Suppose we wanted to live more of our potential as expressions of God. Suppose we wanted to know what color socks God would wear if God were walking around in the material world and needed to wear socks. What would we do? At this point it seems obvious that we need to reduce the fraction of our reality that we view through the lens of fear and increase the fraction of our reality that we view through the lens of connection and love. How do we do that?

The Path To Connection

All of our individual realities are learned. We all had, and have, learning experiences, some based in connection, some based in isolation, where we learned our response to the situations we encounter. We have all done a great job of maintaining those learned responses through many years and often through much contrary evidence. If we believe that we are unworthy of anything good, that the world is dangerous and people are mean and evil, we can certainly find much evidence to support that. But we will also find much evidence that people are good, that nice things can happen to us and the world may have some good places. We humans are fully capable of ignoring evidence that contradicts our learned world view. That is the same mechanism that allows religious people and science people to ignore, deny and repress things that are outside their religious beliefs or their scientific paradigm.

We are all experts in the process of learning and maintaining our realities. I am going to propose here, that we can learn new responses to situations where our original experience was based in fear and isolation. The process of learning and maintaining reality that we have used for our entire lives has worked very well. We could use that tried and true process. This time, however, we understand that the feelings in the learning experience need be love and connection.

If that sounds straightforward, it is somewhat misleading. It appears to be hard to change our learned responses to many situations. Why is it hard? I found the answer to that question in Levine's work with unresolved traumatic experience and its effect on our lives. When our fight or flight response is invoked we are in fear for our lives. When fighting or fleeing is not successful, our entire response is shut down. When that happens in some animals they play dead. That traps a huge amount of energy in our bodies. Animals usually discharge the energy and resolve the traumatic experience. Humans usually do not discharge the energy. The result is a major fear-based learning experience. The response we learn is to recreate the traumatic experience in the hope that we will release the energy and resolve the trauma. Without explicitly releasing the energy it is very difficult to change the learned response.

Why Is It Hard?

Things that we learn in fear usually do not serve us very well and they seem to be hard to change. But some things in our lives are easy to change. What is the difference between things that are hard or easy to change? Keep in mind that the only perception packets that we are interested in changing are those based in fear. The things we learned in love and connection should be serving us well. I believe it is the "strength" of the fear in the original learning experience that determines the difficulty in changing the learned response.

The Learning Spectrum

I generalized that idea into a learning spectrum. A learning spectrum arranges the things that we learn in life according to the strength of the feeling (fear) in the initial learning experience.

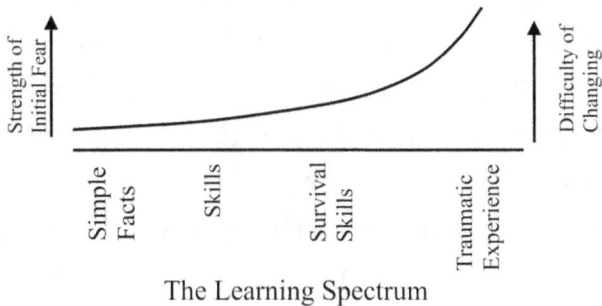

The Learning Spectrum

The items on the spectrum are perception packets. On the left are simple facts about our world, then simple skills, like how to fix breakfast or how to buy groceries, further to the right are habitual behaviors like playing piano or polite behavior in social situations, then come survival skills like recognizing dangerous situations, and on the far right hand side of the spectrum, the behaviors needed to re-create traumatic experiences.

The thing that varies across the spectrum is the strength of feeling (fear is what we are interested in) in the original learning experience. The simple facts we learn may have little or no feeling associated with their learning. Traumatic experiences create the biggest feeling we can have.

We are talking here about changing some of our learned lessons that were originally driven by fear. In thinking about learning new things it is obvious that some things are easier to relearn than others. There appears to be a good correlation between the difficulty we have in replacing an existing experience and the strength of the original fear in the learning experience. The stronger the fear in the initial learning experience, the more difficult it is to relearn or change. I'll call this idea, Doug's theorem, and I offer it here without proof.

Doug's theorem suggests that things on the left end of the spectrum should be easiest to change and things on the right end should be hardest. The theorem accounts, nicely, for the title of Brian Klemmer's book, *If The How-To's Were Enough We Would All Be Skinny Rich And Happy*. All of our efforts to learn new things have to address the feelings that underlie each of the perception packets we are trying to change. Instructions for how to do something new will not help us if we have feelings and beliefs that say we shouldn't do that something. It is very important to recognize that addressing those feelings will take very different kinds of effort depending on the size of those initial feelings of fear.

It is important in our learning process that we address the high-level fears associated with the situation. It also appears to be important that we have some understanding of what the initial feelings were when we learned our original response to the situation. This is because we need to provide feelings of a corresponding magnitude when we learn a new response to the situation.

Body Sensations

The role of body sensations in learning and healing deserves some comment. I have mentioned Levine's work with feelings and Gendlin's felt sense. There is a sub-domain of psychotherapy called somatic psychotherapy. There is even a professional association, the United States Association for Body Psychotherapy. Even with all that, I have the impression, based on my reading and experience with therapists, that somatic therapy is slightly off the beaten path of mainstream psychotherapy. That's unfortunate for some of these reasons.

Our learning process needs to address the low
level, abstract sensations in the body. I think
that Gendlin's work on Focusing demonstrates
why physical sensations are important. Gendlin
found that people who invoke the abstract
physical sensations in their body were
successful in making changes to their old,
learned patterns. People who did not invoke the
physical sensations were unable to make
significant changes in their existing behavior.
This tells us that physical sensations in the
body play a very large role in our learning,
healing, and transformation.

There seems to be what I would call a parallel
thread in more conventional psychotherapy
research. This work is being done under the
banner of interoception, which is defined as the
connection between physical sensations or
awareness and our emotional state. An article in
the April, 2015 issue of *Psychology Today* on
interoception is titled, "Inward Bound". The
status of somatic therapy in the conventional
therapy world is indicated by the subtitle of the
article: "The mysterious connection between
emotion and physical sensation is coming into
sharper focus."

Some work reported by Vivien Ainley and Manos
Tsakiris, indicates the nature of this line of

research. The article is titled, "Body Conscious? Interoceptive Awareness, Measured by Heartbeat Perception, Is Negatively Correlated with Self-Objectification", PLOS l one, http://journals. Plos.org/plosone/article?id=10.1371/journal.po ne.0055568, Feb 6, 2013. They found that women who could not hear or feel their own heartbeat were characterized by self-objectification, which is related to conditions like eating disorders. The general conclusions coming out of this kind of work are that people who are not aware of sensations in their own bodies are likely to have emotional disorders.

Another reason for emphasizing the importance of physical sensation is that sensations in our body are often easier to access than deeply buried fears and beliefs. They are a very good starting place for growing awareness of what's going on in our inner lives.

If we are going to change those parts of our realities that isolate us from all that is, we need to deal with our fear-based learned experiences at all levels, from low-level body sensations up to meaning and physical responses.

Our Culture's Contribution

Changing our fear-based learned responses is a great idea, but it's hard because we have to deal with our large fears, as we just discussed. The task is made even harder because our culture is working against us.

Addressing our hidden fears is a terrible idea for most of us in Western culture. We pride ourselves on being rational and in control. We don't like to appear weak and wimpy. We don't want to admit to being afraid of things that aren't really there. And we don't want to go poking around in our material bodies for feelings and sensations that might not be very pleasant. Our path to serenity and expressing our God-like nature is blocked by many taboos in our culture of origin.

Our patriarchal culture has stereotypical gender roles for everyone. Men don't do feelings. Women don't do power. This is changing, of course. Men are taking care of children. Women are heads of state of powerful countries. But there remain vast stretches of our culture where the stereotypes are still in force. There, the penalties are severe for people who don't play their roles correctly. This is something we learn to be afraid of. That fear causes most of us to bury those parts of ourselves that don't fit the prescribed roles. Jungians call that buried baggage our shadow. We buried it in the first place because it was unacceptable behavior. Trying to bring it back out into the light is still bad behavior. This makes it hard to jump in and explore our dark, inner fears.

There is yet another layer of difficulty, here. Our patriarchal culture expects its citizens to be appropriately respectful and obedient to the male God figure and to His local (male) representatives. I have suggested that we are trying to find our direct connection to God, with no male intermediaries required. We will find plenty of resistance to the idea that we, as mere individuals, can aspire to any sort of direct connection to a God that is bigger than maleness. It's a very interesting journey.

Wearing The Socks

Here is our amazing condition in the material world. We are all direct expressions of the universal intelligence that is God. We have powers that are similar to the powers we usually associate with God. The things we can actually express are limited, of course, by our material form and by our learned experience. Nonetheless, we are all capable of truly amazing things. We all learned our realities. And what we learn can be hell and useful or not healthy and harmful. In either case , in all cases, we all have the ability and the power to implement our views of reality. We are all experts at making our experience exactly what we believe our learned experience tells us it should be.

Against that background, we have the problem of fear. When our initial learning experiences are based on fear, we learn an unhealthy and even harmful view of reality. Feeling the fear distorts our reality and makes it difficult for us to perceive the problem. The closest we can come to perceiving the "truth" about reality lies in our ability to sense what feels right. And fear distorts our ability to feel what's right.

If we want to become healthier (I wouldn't want to say be fixed) we need to replace our fear-based learned experiences with learning experiences based on healthier feelings. This learning process is complicated by the fact that the degree of difficulty associated with learning something new is directly related to the magnitude of the fear in our original learning experience for that situation; the stronger the initial fear, the stronger the feeling for the new experience needs to be.

A Personal Guide Book

The question is, how do we get from here to there? How do we learn to know what color socks God would wear?

Now I am terribly fond of telling other people how to solve their problems. Have a problem? Just ask me and I will give you all kinds of great advice. The problem is that it seems that all the advice I give is really advice for me. I have a hard time figuring out what I should do for my problems, but it's easy to tell other people what to do for their problems. Now, when I feel the urge to tell someone what to do to solve their problems, I make note of my ideas and read them as ideas for me. Even with that awareness, when I talk about things to do, it still comes across as advice for the readers of this document and that is not what I intend. Nonetheless, there are some things that appear to me to be basic in any program of growth and healing. These things arise from the understanding of how things work that I have described in this short meditation, and in my earlier books.

Here are some suggestions for elements to include in any practice for growth, healing and learning.

Safety, Awareness and Acceptance

The first step is to begin where you are. I've always found that to be a comforting idea. I'm already here, where I am. I don't have to do anything or go anyplace to be where I am. It should be an easy place to start.

The second step is to learn how to be aware and accepting of what is. That is very old advice, of course. It is an integral part of many spiritual practices. It sounds like it should be easy since you don't have to change anything about what is. You just have to be aware of it. It turns out to be not quite as easy as that sounds. Being aware and accepting of all that is is a useful definition of enlightenment. And achieving enlightenment usually takes a great deal of work.

I can extend my simple generalizations about fear to account for why being aware and accepting of "all that is" is hard to do. I can define acceptance as awareness without fear. Fear distorts our perception of reality, or at least encourages unhealthy perceptions of reality.

Allowing "everything" into our awareness requires that we get around our fears that are keeping so much of our reality out of our awareness. In spite of that difficulty, I believe that learning to be aware and accepting is an easier task than trying to change our learned perceptions that are based in serious fear.

I learned something from my therapist, that came as something of a surprise to me. She said that it was important to do my healing and transformation work from a place that felt safe. What an idea! I always thought that things had to be tough. I had to tough it out. It also made me realize that there were very few times in my life where I was aware of feeling safe. That was a very interesting lesson for me and I assume it applies equally well to everyone else.

I have been aware of the importance of awareness and acceptance in healing and transformation for quite some time. I have tried several techniques for becoming aware and accepting myself. It has been difficult. I can try to tell myself that I can perceive things without judgement, but the judgement seems very well entrenched in my perception packets. I was very happy to learn of a physical means of feeling accepted. I learned this from my Somatic Experiencing therapist.

This actually began with a practice to feel safe. I realized that one of the rare times when I could say that I felt safe was when I was I bed, with the covers pulled up. It was comforting when my therapist invited me to hold a pillow across my chest and lap and put a blanket over my head. She also had a pillow filled with beans, which made it heavy. She put that across my knees and then across my feet. I found that when I was wrapped up like that, enclosed in soft, warm things that I felt safe. After a little discussion it became apparent that feeling safe is the same as not being judged, which is the same as being accepted.

Another amazing idea. I could feel physically accepted with no mental or intellectual effort at all. It was a pure physical, body sensation. In the sections that follow I will describe ways to invoke and experience feelings in general and fear in particular. The challenge for me and people like me is to become comfortable with those feelings, to accept them. Here was a way to do that. If I sat in my warm and accepting cocoon while I invoked those feelings, I could feel that I was accepting those feelings without judgement. I highly recommend that practice.

Use The Body Sensations

I think it is very important that any practice involves awareness of sensation in the body, of feelings. Here I mean the full spectrum of feelings, from the abstract low-level sensations in the body up to feelings that are emotions and convey information about what feels right, that is, the full content of our perception packets. This idea is both important and difficult. Awareness of sensation in the body is a necessary prerequisite for all healing, growth and change. It is also difficult for a number of reasons. Those reasons include blocks to that awareness caused by our fear, and by our culture which tends to ignore and deny information in the body. What that means for our practice is that some effort may be required initially to develop our lost awareness of body sensation. We have to peel back some layers of learned fear that prevent us from connecting with the wisdom of our bodies. Being wrapped in a cocoon of warm blankets provides lots of skin sensations to help bring your attention to your body.

Like Your Feelings

Another important element of any practice is dealing with feelings, specifically the higher-level feelings that have meaning. The feelings in our new learning experiences have to be of a comparable strength to the feelings in the original learning experiences. That means we might be dealing with strong feelings, and even emotions. This is a serious issue for all of us who are head people, but it can also pose a threat to people who are nominally comfortable with feelings and emotions. Most everyone is uncomfortable with the fears that drove many of their initial learning experiences.

Dealing With Feelings

An important question in deciding what kind of practice to use, is how to deal with strong feelings and emotions, both the old fears and the new, healthier feelings. I said earlier that I seem to be adept at telling other people what to do when I'm really giving myself advice. I've been writing books and giving workshops for a while now and many of them contain things to do to help make us all feel better. As a result I now have a toolbox of techniques that I was sure would be wonderfully helpful for everybody else's problems. And a few people might even find some of the techniques useful for themselves. But what I've really built is a toolbox of techniques for me to use.

Sentics

As a certified head person, any techniques that I would consider using need to protect us from actual emoting. As I look in my toolbox of techniques I find a very safe way to express feelings, called Sentics. Sentics was developed by Manfred Clynes. Clynes found that you can use your fingers to express feeling and emotions by putting them on a solid surface and moving them around. He found that people who did this finger exercise showed the same physiological changes that people who actually felt and expressed those feelings and emotions.

I also know that different tactile experiences can affect how we learn. That is, touching different textures and surfaces can help us learn new things. While I was thinking about all this the idea popped up that I could combine Sentics and tactile experience. I envisioned little squares of wood, each with a different kind of texture. To allow the movement I decided that I would use foam rubber between the block of wood and whatever the textured surface was. I could also allow movement by putting a smooth surface on the block of wood and then putting fabric over it that was loose. That way I could slide my fingers around a little bit on the on the block. I made up a big set of blocks using two kinds of fabric, a shiny smooth fabric and a terry cloth fabric and several kinds of sandpaper.

I found that using these little blocks and my hands to express feelings and emotions worked well. I can visit the feelings and emotions and they feel quite safe since they are at arm's length. One of the features, and the problem, of this technique is that the feelings do feel somewhat remote. That's good for a gentle introduction, but one of the issues that we are dealing with in this process is that we need to be able to handle varying strengths of emotions and feelings. Is there something else I can do?

Puppets

I looked in my toolbox and I found that a couple of years ago I developed and delivered a puppet workshop. I recognized that puppets have been used for therapeutic purposes for a long time. The appeal of puppets is that we quickly get into the idea that the puppet is separate from us. The puppet can do things, say things and feel things that would be uncomfortable for us, but the puppet doesn't mind. I made up a big batch of puppet kits and did a couple of workshops where people made puppets to populate and describe their inner state and their inner issues. The workshops worked very well and people seem to like them.

Most of the puppets were simple sticks with a little circle for the head and a name tag. People could draw faces and make name tags and stick them on the puppet. There was one puppet that was different however. It was the feeling puppet.

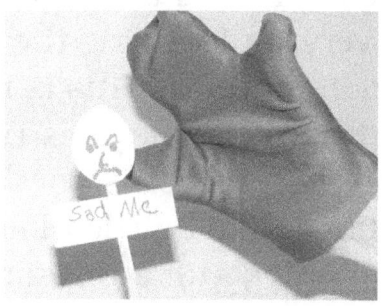

There was only one of these and it was made out of a silky purple cloth. It was a hand puppet roughly in the shape of Casper the ghost. Your hands can be very expressive so it was easy to have the feelings puppet express a range of feelings and different levels of feeling.

Puppets turned out to be more expressive than the Sentics blocks. I could safely express big feelings and little feelings, fear and healthy feelings. The puppets are a good second step in our journey to feelings.

Just Me

Having met with some success in using Sentics and puppets to safely express feelings and emotions, I was feeling a little daring. I thought I might try letting the feelings get a little closer.

The idea is to use my bare hands and arm motions to express feelings. After that idea occurred to me I remembered that I have given a workshop on that very topic, using the hands to express feelings and emotions. This is riskier of course, but I find it manageable. It is riskier because there is no pretense of "something else" expressing the feelings, like a little block, or puppets. It is just me. It's still a physical movement, and I don't necessarily have to do anything with actual inner feelings, but it's getting close.

I found that by using my hands to express both fear-related things and healthier feelings that it is easy to express the size, or strength, of the feelings. The strength of the feeling is related to the amount of muscle tension I use in making the movements both in my hands and in my core.

I've said that the "strength" of the feeling in the initial learning experience determines the "strength" of feeling we need in learning the new experience. How does one determine the strength of the original feeling? I understand that is a trivial question for people who are used to dealing with feelings. But it is a serious issue for those of us who have avoided feelings and expressing emotions for most of our lives. I have found that if I conjure up my feelings of fear around some situation and then act out the feeling with my hands, I can try it several times with varying levels of muscle tension. One level or another will feel right (there is that feeling right thing again). Big fears, for example, seem to want strong muscle sensations. Then I know that when I do the new learning experience with the new feeling I will need something with comparable muscle sensation.

Putting It All Together

Dealing with low level feelings is a necessary, but not sufficient part of any practice that will reduce the role that fear plays in our lives. We need to address all the parts of our perception packets. I have encountered two practices that implement several of the approaches I have recommended, and they both have very good records of producing the desired transformations. I'll describe each practice briefly and then suggest some additions drawn from the understanding we have developed here.

The first practice is called, Integral Transformative Practice. It is described in the book by Leonard and Murphy, *The Life We Are Given*, Tarcher, 1995. The book reports on the experience of two groups that Leonard and Murphy convened. The purpose of the groups was to make changes in selected aspects of the participants' lives. Each participant was invited to set three goals. The three goals were described as normal, exceptional, and metanormal. Normal goals were changes that produced some measurable physical change that is considered a routine kind of change. Exceptional changes were things that did not defy conventional scientific explanation, but would be more than what is normally considered possible. Metanormal changes were things that would be difficult to explain with conventional scientific understanding.

The practice involved simple movement, meditation, affirmations and visualizations. The results were impressive. Seventy six percent of the participants who stayed in the program for six months to a year achieved all three goals.

The ITP process consists of these steps.

The process begins with setting the three goals that are normal, exceptional and metanormal. The next step is movement. The movements are drawn from yoga and Chi Gong. They promote flexibility, balance and body awareness. The last activity is deep relaxation. This is important preparation for the next step, which is transformational imaging.

The active work of learning new realities takes place in transformational imaging. Here, you visualize the desired new state in as much detail as possible. They call for using your hands to direct your attention and intention to the area of the body you are changing. For psychological and spiritual transformations, the hands are placed over the appropriate energy centers. The last activity in this step is relaxation and letting go of the outcomes. This is an acceptance practice.

The final step is meditation. They suggest being aware of your inner activity: thoughts, words, images, feelings. They also suggest becoming aware of a presence beyond the self.

The second practice is by Joe Dispenza which he describes in his book, *You Are The Placebo,* Hay House, 2014. He healed his own smashed vertebrae without surgery. He then went on to study the amazing things that people do when taking placebos. From that work developed a practice that allows people to heal themselves by being their own placebo. The practice that he describes is very similar to the Leonard and Murphy practice. Throughout his book he carefully measures and documents the physical changes that result from the practice.

His process consists entirely of guided visualization.

The first step is induction, which is relaxation and centering. The second step is "finding the present." He accomplishes this by inviting you to become pure awareness, without body or physical senses or material presence.

His third step is changing perceptions and beliefs. He uses a guided visualization where he asks you to change the beliefs associated with your current state and to adopt new beliefs that support our desired new state. He emphasizes being aware of the feelings associated with both the old beliefs and the new beliefs. The feelings connected with the new beliefs need to be stronger than the feelings connected with the old beliefs.

Some Suggestions

With the understanding we have developed up to this point, I think we can make these practices more effective with these two suggestions.

Awareness of body sensations associated with feelings and emotions is crucial to any successful transformation or healing. We could add some work at the beginning of each practice to explore current feelings and body sensations. Gendlin would describe it as finding the felt sense. Exploring your current feelings, no matter how negative, is an awareness and acceptance practice, which I believe is a necessary first step for all healing

I think it is important to use a physical technique to express the current and new feelings and emotions. I suggest Sentics, puppets or hand motions. It is important to involve the body in all transformation work.

It would help to do the visualizations while safely wrapped in a blanket cocoon. The feelings for the new experience should be from the love side of the spectrum.

That all sounds very promising, and it is. All aspects of our beings are malleable. But, for all of us who are really seeking quick and easy transformations, here is a sobering reality check. Both *Placebo* and *Life We Are Given* agree that it takes six months and more of practice to make physical changes. Our learned realities are deeply learned, which means that changing them is likely to take comparable effort. Any practice will take time to produce results, no matter what the self-help books promise.

Dualities, Again

I have devoted considerable effort in this small volume to show that things we think of as being poles apart, as being dualities, are really parts of a single entity. I described the two sides of the duality as being partners in a dance where there are no dancers, only the dance. Now I seem to have introduced another duality, two things that are different from each other: separation and connection, love and fear. These seem to be different. They produce very different results when they drive our learning experiences. But are they really different? And if they aren't different, can we eliminate the "bad" one and just embrace love?

As I understand it, there is no separation in the field. When we are only in the field, that is when we don't have a material body, it appears that we are aware of the infinite connection. Evidence for that idea comes from the many reports of near death experiences. That means that separation and isolation are phenomena of the material world. Here in the material world we need differences in order to perceive anything. In order to perceive connection, we need to have separation to allow us to see the connection. We need fear to be able to recognize love. Here in the material world, love and fear are connected. You can't have one without the other.

The Taoists are big on that idea and they have a very famous symbol to represent that situation.

The yin and yang symbol represents two opposites that are intimately bound together. Here in the material world, love and fear are the two sides of a yin and yang picture. In the non-material world there is only connection, or love.

The duality of love and fear appears to be necessary for our perception here in the material world. It does not appear to be necessary for conscious perception when we have no material body, that is, when we are only "in" the field, or in heaven if you prefer to think about it in those terms.

Our wave functions persist across lifetimes of our material bodies. This implies that some of our learned experiences should persist, as well. When we start the next material lifetime we can have some learning from previous lifetimes. This is another old, spiritual idea called karma.

We move between the state of having a material body and not having a material body. When we have a material body we can perceive the world through our learned fears and separations. When our material body dies, we enter a state where our perception is entirely open. That would account for the life review that is part of most near death experiences. We are aware of everything without judgement in our non-material state and we go back to dealing with our learned fears and isolations when we return to the material state.

Love and fear really are different things, but they are dancers in the intimate dance of perception here in the material world.

What Color Socks Does God Wear?

I didn't like organized religion from a young age, but I thought there was value in some kind of spirit. I spent my life looking for spirit in the places where I was comfortable, in science and in science fiction. And now I have found a way to explain how the world works in scientific terms that includes many of the old, persistent spiritual ideas, like heaven, soul and God. In this view everything is a direct expression of the universal consciousness of the quantum field. We are each animated by that consciousness all the time. We are always connected to the field and to all that is in the field.

While these ideas support many old teachings, they also conflict with many teachings of organized religions. I will confess to finding considerable satisfaction in that. My sense of justice and balance is pleased that in addition to contradicting some of the rules of organized religion, this world view contradicts some of the cherished rules of mainstream science.

On the religion side there is no separation between God and human. There is not judgement required for us to enter heaven because we are in heaven all the time. It is the only place we can be. On the science side, many of the things science has ignored and rejected since Descartes turn out to be real, normal phenomena in our world. There is a large scale, creative (but probabilistic) consciousness at the ground of our universe. The paranormal phenomena of knowing and influencing things at a distance are real and normal parts of all life. We do have a non-material essence that persists across our material lifetimes.

I find all that very appealing, very promising. Here is a view of spirit that I can describe in the language of science. It supports most of the basic (mystical) truths that sages have taught for 3,000 years. The socks we wear can be God's socks.

We can all live like we are expressions of the highest intelligence in the universe.

Now that I understand what is possible, I would like to help all those people out there who struggle with less that God-like lives. That sounds very altruistic. But I learned fairly late in life that the problems I see in other people and the (very good) advice I give them to fix their problems are really my problems and advice for me to fix my problems. Fortunately for some of the readers of this book, my problems are hardly unique. My advice for me might actually benefit other people.

My intellectual/science approach to spirit and healing led me to (surprise, surprise) something that has been a problem in my life: feelings are really important in all aspects of our lives and are the basis of our connection to the consciousness of the field. A problem in my only-slightly-less-than-optimum life is my avoidance of feeling and emotions in my conscious life. I acquired this trait from my only-slightly-less-than-optimum parents, who were experts at avoiding feelings and emotions. This accounts for the strange combination of concerns that I address in this and my other books. One concern is a scientific explanation of what is usually expressed as feeling, that is, God and sprit. The other concern is elevating feeling, and especially fear, to the main obstacle and main enabler of connection with the field, that is, with God.

Our perception mechanism is the connection between the lofty view of spirit and the very mundane issue of fear in our learning experiences. The human characteristic of learning our realities makes this mechanism very important in our lives. The result is that when we learn something in fear, what we learn is separation and isolation. And this leads to all of the decidedly un-God-like behavior we see in the world. This is both alarming and very promising.

It is alarming because all of my troubles are "my fault", to put it negatively. Of course, all my strengths and positive contributions are my responsibility, too, which I am happy to take credit for. It is much easier when my troubles are someone else's fault, which is why blaming someone else is so popular.

It is promising because if I learned my realities once, then I can learn them again, differently. We know that is possible because there have always been a few people who have made large changes to their learned realties, from being poor to being wealthy, from being sick to being healthy, and in my case, from being afraid of public speaking to being very comfortable giving talks to groups.

While I know all that and I can explain how it works at length, it still seems to be hard for me to move past much of my fear-based perception packets. That has been a problem in my life. As I look around I see many other people who seem to be stuck in realities that are not serving them. Some are even aware of their own role in making those realities and they still find it hard to change.

This is why I feel compelled to include words about practice for changing our realities in this and my other books. I like to think that the difficulty that I have had in moving past my learned fears makes me a good (or, at least, better) judge of practices to help us on our paths.

For me, a certified head person, understanding how the world works and how I work is a necessary first step on the path to becoming a whole person (with feelings). I find it encouraging to know what is possible. I find it especially encouraging to understand that I am part of all that is. I am connected to a direct expression of the consciousness that people have always called God. And all that connection remains in place no matter what fear-based thing I have learned about the world and my place in it.

Message to self, and anyone else in a similar position: Serenity surrounds you. Do the work so you can feel the peace, and know what color socks God wears.